Perspectives from Former Executives of the DOD Corporate Research Laboratories

By Richard Chait

Center for Technology and National Security Policy

National Defense University

March 2009

Richard Chait is a Distinguished Res earch Fellow at CTNSP. He was previously C hief Scientist, Army Material Comm and, and Director, Army Research and Laboratory Management. Dr. Chait received his PhD in Solid State Science from Syracuse University and a BS degree from Rensselaer Polytechnic Institute.

Acknowledgements

The author expresses deep app reciation to the in terviewees (Drs Timothy Coffey, John Lyons, and Vincent Russo) for their wil lingness to be interviewed and for their candid thoughts regarding their successful careers as leader s of the Service co rporate research laboratories. Many thanks are due to Dr. Lance Davis, who provided the OSD perspective on key laboratory legislative m atters that affected the Serv ice laboratories during the Coffey-Lyons-Russo tim eframe, and to Dr. Tom Killion for his review of the manuscript. Also, the tim e spent by Mr. Jordan W illcox in copyediting the text is acknowledged with appreciation.

Defense & Technology Papers are published by the National Defense University Center for Technology and National Security Policy, Fort Lesley J. McNair, Washington, DC. CTNSP publications are available at http://www ndu.edu/ctnsp/publications html.

Contents

Introduction

Recent studies at the National Defense University (NDU) have documented the important science and technology (S&T) contributions of the m ilitary service laboratories.[1] These studies showed that in-house Departm ent of Defense (DOD) laboratories, in cooperation with the private sector and academia, developed critical technologies for weapon systems that strongly impacted the outcomes of World War II and th e Cold War. Involvement of the in-house laboratories continues today, undimi nished, as our Nation battles the threat of international terrorism.

The fundamental research com ponent of th e in-house laboratory system within DOD rests with the Arm y, Navy, and Air Force cor porate research laborat ories, henceforth called CRLs in this pap er. These are the labo ratories that perform the basic res earch that underpins the S&T programs for each Service. It is the purpose of this paper to document the thoughts and opinions of individuals who have led research activities at each of the following CRLs: the A rmy Research Labor atory (ARL) in Adelphi, MD; the Naval Research Laboratory (NRL) in W ashington, DC; and the Air Force Research Laboratory (AFRL) in Dayton, Ohio respectively. Thes e individuals are: John Lyons (ARL), Timothy Coffey (NRL), and Vincent Russo (AF RL) and are referred to collectively as "former laboratory executives" or "executi ves" below. These executives had over 100 combined years of experience directing labo ratory research. The in terviews with Lyons, Coffey, and Russo occurred during February 2008. The Lyons and Coffey discussions were conducted in-person, while a teleconference was used for the Russo interview.

Directed actions from both the Congress and the Executive Branch have had a significant impact on the CRLs. Some of these action s have been accompanied by studies that have also played significant roles. W hat follows is a discussion of som e of these im portant directed actions and studies, so as to place in perspective both the history of the CRLs and the views of those executives who ran them.

Following the discussio n of directed actions and studies, historical highlights of each CRL will be presented. This will be f ollowed by the biographies and interview highlights of each form er laboratory executive, as well as additional perspectiv es as seen through the eyes of for mer senior members of the O ffice of the Directo r, Defense, Research and Engineering (DDR&E). The paper concludes wi th a discussion of the various viewpoints expressed in the interviews. Included here are recommendations that seek to leverage the vast amount of experience possessed by the former laboratory executives.

---

[1] Richard Chait, John Lyons, and Duncan Long, "Critical Technology Events in the Development of the Abrams Tank: Project Hindsight Revisited," *Defense and Technology Paper 22* (Washington, DC: Center for Technology and National Security Policy, 2005). Also see Richard Chait, John Lyons, and Duncan Long, "Critical Technology Events in the Development of the Apache Helicopter: Project Hindsight Revisited," *Defense and Technology Paper 26* (Washington, DC: Center for Technology and National Security Policy, 2006). Finally see John Lyons, Duncan Long, and Richard Chait, "Critical Technology Events in the Development of the Stinger and Javelin Missile Systems," *Defense and Technology Paper 33* (Washington, DC: Center for Technology and National Security Policy, 2006).

Important Studies and Directed Actions

As noted earlier, the D OD in-house laboratories have played a vital role in our Nation's defense. Given this role and the visibility that comes with it, it is not su rprising that they have been the subject of many studies and/or actions. It has been estimated that about 100 studies and related reviews of governm ent laboratories, also refe rred to as in-house laboratories, have been conducted since 1962. [2] Most have been initiated by Congr ess, The White House, or the Pentagon. It is not possible to provide details of each and every study. However, there are several that have had a marked influence on the for mation and operation of the Service laboratories. Note th at these studies and actions, presented in a generally chronological fashion below, have emphasized consolidation and increased efficiency.

One of the most notable laboratory review s was conducted in 1982. U nder the authority of the White House Science Council, a Fede ral Laboratory Review Panel was appointed for the purpose of reviewing Federal Laborat ories and making recommendations in areas of utilization and perform ance. The Panel' s charge included lookin g at laboratory missions and identifying any system ic impediments to performance. The Panel issued its report, generally referred to as the Packard Report, in 1983. [3] They m ade a num ber of recommendations pertaining to the Federal Laboratories' missions, personnel, faculty, and management. Among the m ost important w ere the adoption of peer review and the empowerment of the laboratory directors in cr itical areas of laboratory management. One of the examples cited in the Packard Re port was the experim ent in personnel management being conducted at the Navy's Surface Weapons Center in China Lake, CA and the Naval Ocean System s Center in San Diego, CA. Known as the China Lake alternative personnel system or simply "C hina Lake Experim ent," this 1980 project included a streamlined personnel classification system based on performance, rather than on longevity. One of the m ost important things the Packard Report did w as to recommend that all Fed eral laboratories could benefit by applying aspects of the China Lake experiment. In 1984, based on a Packard Report progress report, [4] the White House Office of Science and Technology Policy draf ted legislation to enable other federal laboratories to follow up on the Chi na Lake experiment. However, Office of Personnel Management (OPM) and personnel officers in other Cabinet Departm ents resisted, and the legislation did not move forward.

---

[2] Timothy Coffey et al, "Alternative Governance: A Tool for Military Laboratory Reform," *Defense Horizons 34* (Washington, DC: Center for Technology and National Security Policy, November 2003), available at <http://www ndu.edu/inss/DefHor/DH34/DH34.pdf>.
[3] Report of the White House Science Council Federal Laboratory Review Panel (Washington, DC: The White House, May 1983).
[4] Office of Science and Technology Policy, *Progress Report on Implementing the Recommendations of the White House Science Council's Federal Laboratory Review Panel, Vol. 1—Summary Report* (Washington, DC: Office of Science and Technology Policy, July 1984).

A 1987 Defense Science Board report supported a streamlined management process that would provide for effective and effici ent operation of the DOD laboratories. [5] This led to the creation in 1989 of the DOD-wi de Laboratory Demonstration Program (LDP), which sought many of the features contained in th e China Lake experim ent. To m ove things along more quickly, the S&T executives of the three Services established the Laboratory Quality Improvement Program (LQIP) in 1993, as a m eans to re-en ergize the LDP initiative to improve the quality and producti vity of the DOD laboratories. Under LQIP, some DOD laboratories m ade headway by stre amlining their business practices in such areas as civilian personnel, financial management, information infrastructure, contracting, and facilities renewal. The goal was to grant the heads of the DOD laboratories increased authority to choose the m ost cost-effective means of ope rating their organizations. Initiatives included designing and implementing streamlined civilian personnel and R&D contracting procedures; im proving facility renewal efforts, using increased m inor construction thresholds; design ing financial managem ent approaches that allowed the identification and comparison of the true cost of doing busines s; and creating an information infrastructure that wo uld aid th e scientists and engineers in exchanging analytical information.

In line with these in itiatives, the Of fice of the Secretary of De fense (OSD) and the Services established Project Reliance in 1990. The objective of this initiative was to reduce duplication across the Services, and improve coordination and integration. Only the three Services initially participated in Project Reliance; it was subsequently expanded to include other DOD activitie s, including the Defense A dvanced Research Projects Agency and the Ballistic Missile D efense Organization. As participation grew, Project Reliance was expanded and becam e part of a more com prehensive DDR&E develo ped strategy and planning method, which included the following: a Defense Technology Area Plan for presenting the DOD technology inve stment plan and strategies; Defense Technology Objectives to define return on i nvestment; and an independent review called the Technology Area Review and Assessm ent (a "TARA review") to assess integration and recommend opportunities for improved synergy among the Services. These processes resulted in an improved investment strategy for each Service.

The LQIP and Project Relian ce efforts were part of the response to the White House and the Congress as they pushed for m ore emphasis on reduced infrastructure through more cross-service integration and Service laboratory consolidation, all in an ef fort to meet the challenge of reduced R &D funding. In additio n, several actions and m echanisms were used by the Congress and the W hite House. At the top of the list was the DOD -initiated, congressionally approved, Base Realignment and Closure (BRAC) action. Many closures and consolidations have occurred as a result of the BRAC process. For example, the 1991 BRAC disestablished and consolidated management of nine Army laboratories under one command, and led to the creation of the Ar my Research Laboratory. Sim ilarly, the 1993 BRAC and the 1995 BRAC disestablished and tr ansferred functions of the Belvoir R&D Center and Aviation Troop Comma nd. The Na vy and Air Force took sim ilar actions during this tim e period. The Air Force consol idated its laboratories into four large

---

[5] Defense Science Board, *Report on the 1987 Summer Study on Technology Base Management* (Washington, DC: Department of Defense, 1987)

laboratories (called "super" laboratories in Air Force parlan ce), while the Navy consolidated the technical infrastructure of four Warfare Centers.

Despite BRAC, LQIP, and the Project Reliance initiatives, there was a belief within the DOD leadership that m ore reforms were necessary. What the Pentagon had in m ind was embodied in the Defense Management Repor t Decision 922 (DMR D 922) in late 1989. [6] Here, special in-house groups were appointed to investigate options for consolidating DOD functions, including the advantages and disadvantages of interservice and intra-service consolidation of laboratories. As a result, the Servic es were dire cted to explore the entire range of laboratory options, including alternatives to a concurrently considered proposal to create an overarching DOD laboratory.

Additional input resulted from a very extensive study undert aken in 1991 by the Federal Advisory Commission on Consolidation an d Conversion of Defense Research and Development Laboratories. [7] The purpose of the Comm ission, which was established by public law, was to study the DOD laboratory system and provide recommendations to the Pentagon on the feasibility and desirability of various means to improve their operations. In its study, the Comm ission reaffirmed that the laboratories within each service "are a function of that Service's weapon system s acquisition structure" and that "there was no need to force the serv ice laboratory systems into a sing le model". Recognizing the need to improve the effectiveness of the DOD laboratory, there was also strong support for Project Reliance, as well as the laboratory consolidation efforts noted above for the Army, Navy, and Air Force.

With a change in administration came another round of laboratory reviews. In November 1993, the National Science and Technology Council (NSTC) was established with the aim of conducting an in-depth review of laboratories from several federal agencies. This Interagency Federal Laborator y review concentrated on la boratories operated by the DOD, Department of Energy and NASA. A report was issued in May 1995 that concluded that the laboratory systems of these agencies provided essential services to the Nation. [8] While the W hite House en dorsed the report, it noted that the DOD needed to explore the advantages of cross-Service in tegration more thoroughly. To address the cross-Service issue, the Pentagon was directed to issue a follow-up report in early 1996.

Congress also had an interest in improving the efficiency of DOD laboratories. In 1996, it passed the National Defense Authorization Ac t (NDAA). Section 277 of this legislation directed DOD to develop a 5-year plan to set forth specific actions needed to "consolidate the laboratories and test and evaluation centers." To initiate the effort, the Secretary of Defense was instructed to subm it an init ial plan to Congress no later than May 1996,

[6] Department of Defense, *Defense Management Report Decision 922* (Washington, DC: Department of Defense, October 1989).
[7] Federal Advisory Commission on Consolidation and Conversion of Defense Research and Development Laboratories, *Report to the Secretary of Defense* (Washington, DC: Federal Advisory Commission on Consolidation and Conversion of Defense Research and Development Laboratories, September 1991).
[8] National Science and Technology Council, *Report to the White House* (Washington, DC: National Science and Technology Council, May 1995), available at
<http://www.ostp.gov/galleries/NSTC%20Reports/Federal%20Laboratory%20Review%201995.pdf>.

outlining the DOD strategy for acco mplishing the consolidation and restructuring of the laboratories and test centers.

Since there were now two acti ons underway stressing reform—the response to the NSTC and the White House as well as the response to the 1996 NDAA and Congress —it was decided to combine the outcomes from the two studies into a sing le plan called Vision 21.[9] As requested, the report discussed ways to reduce cost, elim inate duplication, and maximize efficiency and effectiveness for the DOD laboratories. The plan identified three key pillars in accomplishing the desired laboratory reform. These were:

- Reduction of infrastructure costs with emphasis on high-maintenance and inefficient facilities while retaining critical capabilities
- Restructuring resulting from improved processes and cross-service reliance
- Revitalization of key laboratories with an emphasis on critical technologies.

In essence, the goal of Vision 21 was to provide a plan for elim inating unnecessary infrastructure, at the sa me time maintaining the research and development program s and facilities essential to developing the technology for weapon systems of the future. As will be discussed in later sections of this paper, Vision 21 played an im portant role in the Air Force's decision to continue to overhaul its laboratory infrastructure, ultimately resulting in the creation of AFRL.

In addition to the 1996 NDAA, there have be en other congressional actions taken to promote DOD laboratory refor m. In Section 913 of the 2000 NDAA, Congress proposed that DOD use university study team s to look at the relevance of the defense laboratories and evaluate their current work and utility in the future. The studies found that relevant work is being performed and that the laborat ories are well focused on the technical n eeds of the services. Som e concerns were noted. These includ ed a heavy concentration on short-term needs at the expens e of longer-term opportunities, and the need to continue to address challenges in the science and engineering (S&E) personnel area.

Congressional actions conti nued to focus on the challenges in the S &E personnel area. For example, wanting to see a faster pace of reform, Congress acted in Section 342 of the 1995 NDAA, to m ake the China L ake experiment perm anent. As a result, m any Service laboratories applied for the laboratory de monstration status, which enabled them to carry out personnel demonstration projects similar to the China Lake experim ent. Laboratories that applied for this st atus and were approved b ecame known as "Reinvention Laboratories" and were overseen by DDR&E under LQIP.

Section 9902 of the 2004 NDAA was also a ve ry important piece of legislation, as it called for a DOD-wide personnel system overhaul. This new personnel system, called the National Security Personnel System (NSPS) , replaced the for mer 15-step General Schedule system with a four-band system , and adopted the principle of pay for performance. As shown in Table 1, the auth orizations offered by NSPS did not provide

---

[9] Department of Defense, *Vision 21—The Plan for 21st Century Laboratories and Test and Evaluation Centers of the Department of Defense* (Washington, DC: Department of Defense, April 1996).

the laboratories with the degree of flexibility provided by the reinvention status under LQIP. It is important to note that one of the main objectives of LQIP was to improve laboratory management by allowing laboratory managers to waive many regulatory statutes. These powers are not available under NSPS. NSPS would appear instead to move most of the decision authority to higher levels rather than to delegate to individual laboratory managers.

Table I
Comparison of Elements of LQIP vs. NSPS[10]

LQIP	NSPS
Can waive many parts of Title 5, the Civil Service System	Cannot waive these items
SECDEF approves changes	OPM must approve changes
May pay starting salaries anywhere in a pay band	Limited to 30 percent above minimum
Has a Pay Band V for senior positions	Does not have such a band
Supervisors not automatically paid more than group members	Assumes supervisors are paid more
Can promote from band to band without competition	Crossing pay bands requires competition
Can manage most of Human Resources (HR) functions	HR functions performed above the laboratory level
Classification, recruiting, qualification, and hiring authorities reside with laboratory managers	Classification and related actions are performed above the laboratory level

It should be noted that in approving NSPS, Congress exempted the Service laboratories until 2013 to allow for an orderly transition. Congress also revisited some S&E personnel issues in the 2005 NDAA (Section 1107) and 2006 NDAA (Section 1128). These legislative actions requested studies comparing NSPS and LQIP to each other as well as to systems utilized in the private sector. To date, these studies have not been completed. With this background in place, the next several sections present the perspectives from former CRL executives.

---

[10] William McCorkle et al., memorandum to W.S. Rees, Chair, Laboratory Quality Enhancement Program, Office of Director, Defense Research and Engineering, August 2006.

6

John Lyons and the Army Research Laboratory

Prior to the formation of ARL, the Army had research facilities at several locations. One, the facility at Watertown Arsenal, where pyrotechnics and waterproof paper cartridges were studied, dates back to 1820. As the Army's technology needs grew, other laboratories had been formed. Among them were the Atmospherics Science Laboratory (White Sands, NM), Harry Diamond Laboratory (Adelphi, MD), Human Engineering Laboratory (Aberdeen, MD), Vulnerability Assessment Laboratory (White Sands, NM), Ballistic Research Laboratory (Aberdeen, MD), and Electronic Devices Technology Laboratory (Ft Monmouth, NJ). Over time, the laboratory located in Watertown, MA evolved into the Materials Technology Laboratory.

These laboratories operated independently, each reporting to the Army Materiel Command (AMC) in Alexandria, VA. This reporting chain was changed in 1989, when Laboratory Command was formed under AMC. The laboratories were now managed by a single reporting element commanded by a general officer. Several additional actions taken created further organizational changes. The most important of these was the previously discussed 1991 BRAC action, which established ARL.

Biography

John Lyons has a varied background, including an undergraduate degree in chemistry and a PhD in physical chemistry, as well as extensive experience in both the private and public sectors. He served with Monsanto Company for 18 years in various research and development positions, starting at the research bench and then holding various positions in management of research and development. Following his stay at Monsanto, Lyons joined the Department of Commerce's National Bureau of Standards (today known as the National Institute for Standards and Technology or NIST) in Gaithersburg, MD. At NIST, he was at first Director, Center for Fire Research; and then Director, National Engineering Laboratory; before being appointed by the President to serve as Director, NIST. In total, Lyons spent 20 years at NIST. Following his career at NIST, Lyons moved to the Department of Defense as Director of the Army Research Laboratory. He served in that position for over 5 years before retiring from government service in 1998. In summary, Lyons spent a total of 18 years in the private sector and 25 years in the public sector. He is now a Distinguished Research Fellow at the National Defense University's Center for Technology and National Security Policy.

Interview Highlights

Lyons' approach to managing ARL was influenced by his private sector experience. At Monsanto, he observed that those individuals who took a fundamental approach to solving industrial problems and were able to publish the results of their research in refereed journals often had successful careers, whether in research or management. His appreciation was established there for the important role of basic research within both the private and public sectors. The thought here was that the grounding one gets at the research bench provides the basis for the sound technical judgment that is critical to a successful career.

At NIST, Lyons was able to apply what he ha d learned at Monsanto. He observed that in its content, the Center for Fire Research program was more of an applied engineering program than a research program. Believing that some basic research was needed, Lyons created a fire science program , forming the underpinnings of what became a nationally recognized activity at NIST . The National Science F oundation (NSF) was funding a similar fire science program in academia. Because of the upgraded NIS T program and a change in emphasis at NSF, Congress transfe rred the NSF program to NIST. As a result, NIST had both a rapidly i mproving in-house technical program and a first-rate academic effort. The NIST fire research program soon gained a reputation as a world-class fire research capability.

Having gained this reputation, it was important to Lyons for NIST to recruit and keep top scientists and engineers. For exa mple, one NIST scientist, who was to go on to win a Nobel Prize, was heavily recru ited by a l eading university. NIST management provided strong support for his research, excellent la boratory equipment and fa cilities, and the freedom to pursue good science. This support was im portant in NIST's ability to retain this scientist. As evidence of this emphasis on basic res earch and quality work, NIST scientists have won three Nobel prizes since 1997.

Lyons also found the potential interactions and funding streams at NIST quite interesting. In contrast to what he w as going to encount er at ARL, funding was appropriated directly to NIST by Congress, and L yons testified bef ore the aut horization and appropriation committees. Congress was also interested in maintaining contact with the N IST leadership, since they had been ins trumental in reorganizing NBS into NIST. It was thus easy for Ly ons to establish a line of co mmunication with Congressional committees. Another NIST observation was that the cust omer base was different; it was m ade up largely of scientists and engi neers in academia and industry. NIST's importance to these customers stemmed from NIST's goal to es tablish the technical stan dards needed to support the increasingly sophist icated work in science, e ngineering, and manufacturing. As a result, Lyons noted a very positive attit ude on the part of the private sector rather than the less than pos itive comments that are sometimes heard regarding the governm ent S&T or technology development laboratories.

At ARL, Lyons found certain di fferences in comparison to NI ST. At NIST, he had been free to talk with congressional staffers, resul ting in frequent interactions. At ARL on the other hand, such interactions were less frequent, as they we re handled by higher headquarters. In this respect, ARL was m ore like Monsanto. W hile the interface with Congress was different at ARL, Lyons' m anagement techniques were similar to those he used at NIST. He was protective of basi c research funding, beli eving, as he had at Monsanto, that a sound underpinning of funda mental research was of great benefit to the technology development programs. He believed that this sound research base, coupled with an excellent technical staff, led to an excellent scientific product. ARL was thus in a solid position to be an honest broker in the Army acquisition cycle, should the opportunity arise.

The technical decisions Lyons made were based on both his interaction with his technical staff and "gut feel". His com fort with th e technical staff stemmed in part from a management style that got him "out and ar ound" both internally and externally. Lyons frequently used the outside technical community to benchm ark his program s. For example, at NIST he gained from visits to many public and private-sector laboratories. One example was his visit to Bell Laborat ories to discuss his fiber optics progra m initiatives. At NIST, the National Research Council (NRC) had provided peer review of the technical quality of the program s. This practice was continued at ARL, where he initiated an ARL Technology Assessm ent Board (TAB) with the NRC. The TAB e ffort reviewed the technical program s of each ARL directorate, covering key areas such as sensors, armor, anti-arm or and human factor s. He found the TAB critiques very helpful, and was diligent in following up on TAB reco mmendations. Lyons was m indful of what it took to have a wor ld-class laboratory, having served on th e Federal Advisory Committee mentioned on page four. He used sel ected metrics (patents, publications, etc) to monitor ARL's response to the TAB's recommendations. In additio n to the oversig ht of technical quality, Lyons utilized an Army Materiel Command Boa rd of Directors (technical directors of other AMC laboratorie s) and a group of stakeholders drawn from the Army's senior leadership, the latter to address overarching relevancy and policy matters.

Lyons also brought about increased interacti on with the private se ctor, using partnering initiatives that leveraged i ndustry's and academ ia's acknowledged strengths in given technology areas. This Federa ted Laboratory Initia tive, as it was originally known, formed a competitiv e selection of external laboratories into consortia, in order to work jointly with ARL to m eet the Arm y's expanding research needs in areas such as telecommunications, simulation, displays, and advanced sensors. The concept was to rapidly secure advanced technologies in areas where ARL had not previously concentrated, while at the sam e time building up internal com petence at ARL. He contrasted this approach to that of simply contracting out the work. This highly successful concept continues today as the Collaborative Technology Alliances.

Lyons expressed pride in th e above accomplishments, and believes that they have contributed significantly to ARL's growing reputation in the technical community.

Timothy Coffey and the Naval Research Laboratory

Established in 1923, The Naval Re search Laboratory (NRL) is the oldest of the m ilitary service laboratories. It was opened at the instigation of Thomas Edison, who wrote in the New York Times that "the governm ent should maintain a great research laboratory—in this could be developed all the techniques of military and naval progression without any vast expense." NRL's original organization consisted of three divisions: Radio, Sound, and Metallurgy. The laboratory did pioneering work in the fields of radar, high frequency radio, underwater sound propagation, and def   ect analysis in m etal castings and weldments. During WWII, the staff grew near ly ten-fold, and five new NRL divisions were added in the areas of optics, chem istry, metallurgy, mechanics and electricity. T he expanded research effort produced important devices and s ystems for t he war effort in areas such as radar, son ar, and cou ntermeasure systems. Also, a new therm al diffusion process for separating the U–235 isotope was developed in support of the Manhattan Project to develop the first atomic bombs.

Following WWII, NRL was able to adopt a   long-term perspective and address basic research issues associated with the opera tional environment: earth, sea, sky, and sp ace. To reshape and coordinate the research, NR L needed to transform a group of largely autonomous scientific division s into a unified organizatio n with a clear m ission and a fully coordinated research program . The first attempt at reorganization vested power in an executive committee composed of all the di vision heads. Further streamlining came in 1949, when a civilian Director of Re search was named and given full authority over the program, which today includes most of the science and technology areas of interest to the Department of the Navy. Afte r WWII, NRL pioneered naval research that led to the Nation's first intelligence satellite, Global Positioning System, and m olecular structure analysis, for which two NRL scientists received the Nobel Prize.

Biography

Coffey's background includes an undergraduate degree in electrical engineering and a PhD in physics. Following graduate school, he spent several years with EG & G Inc. as a research consultant working projects involvi ng theoretical and mathematical physics. He became familiar with N RL programs during this time period and as a result dec ided to join the NRL staff. He spent the next 30 years at NRL, beginning as head of the P lasma Dynamics Branch, where he d irected research in areas such as the s imulation of plasma instabilities and the developm ent of computer codes for chem ically reactive flows. His next positions at NRL were as head of the Plasma Physics Division, Associate Director of Research, and then Director of Research. He   occupied the latter position for 20 years before retiring from government service in 2001. Currently, Coffey is a Distinguished Research Fellow at the Center for Technol   ogy and National Security Policy at the National Defense University.

Interview Highlights

Like John Lyons, Coffe y's strong research background had a m arked influence on his vision for NRL. He believed that N RL should be a "bona-fide card-carrying m ember of the scientific and techn ical community" and that it was his job to m aintain NRL's scientific and technical reputation as a major research laboratory. His approach to making the critical scientific and technical decisions involved in this task required him to gather input from a number of sources. These included in-house technical discussions, where he would hear from the proponents of competing scientific and techni cal approaches. He would also seek opinions on these approaches from outside experts. In addition, he would attend scientific and technical conferences and symposia to hear m embers of the NRL scientific and technical staff present papers . This provided him with an opportunity to evaluate the presentation, as well as hear the comm ents and questions that took place during the session. After attending a num ber of these sessions, Coffey was able to draw conclusions related to proposed research or the quality of ongoing NRL research.

The emphasis on scientific and technical qua lity also was apparent when it cam e to making decisions regarding selections for heads of NRL divisions and branches. His priority was to se lect the person with the be st technical qualificati ons. He believed that these people were paid for making the sound technical decisions and to provide scientific and technical leadership. The m ore administrative and procedural matters could be handled by a well-trained office staff.

Because Coffey believed that the NRL scientists needed to be visible in the scientific and technical community, he tended to monitor their activities, using metrics such as papers, presentations, patents, and peer recognition that stressed scientific and technical output. He used peer reviews to judge perfor mance of NRL program s. Early in his tenure as Director of Research, Coffey contracted with the Nationa l Research Counc il to assist with the re view of the NRL program . He ultimately settled on the use of ad-hoc committees of scientific and technic al experts in the areas being reviewed. Mem bers of the committees were drawn from industry a nd academia, and were nationally recognized as subject matter experts in the appropriate fields.

To answer the question of whether an S&T la boratory is world class or not, one must examine its performance in com parison to the res t of the sc ientific and te chnical community. Coffey believes that NRL passed this tes t with high marks. Maintaining the world class reputation of NRL not only involved paying close attention to the quality and importance of the scientific a nd technical programs as noted above, but also to attracting and retaining top quality scien tists and engi neers. For this to be realized, NRL placed emphasis on quality of life factors (equipm ent, facilities, su pport services, etc) and on providing competitive salaries to c omplement the em phasis on scientif ic and tech nical quality.

Among his successes, Coffey cites NRL's consis tent support from the Navy's scientific and technology base, as well as from the acquisition and development community. Also, NRL maintained its s tatus as one of the top DOD laborator ies, as reflected by its technical achievements, new program s, quality of the technica l staff, patents, and publications. Coffey's disappointm ents include the feeling that despite these successes, NRL seemed to be in a constant "surviva l drill". Yesterday's accomplishments and contributions were dismissed with the questi on, "but what have you done for m e lately?"

It became clear that th ere was no way to "win;" rather one had to ensure that NRL "did not lose."

He was also concerned about the increasing outsourcing of the S&T that was traditionally performed by the DOD in-house laboratories . In addition, the growing dependence on non-government entities in im portant aspects of S&T adm inistration was bothersom e. Coffey's thoughts on this subject will be dis cussed in more detail in th e Discussion and Concluding Remarks section of the paper.

Vincent Russo and the Air Force Research Laboratory

The Air Force Research Laboratory (AFRL) was activated in 1997. Prior to the creation of AFRL, the Air Force conducted its research at four major laboratories containing thirteen different operational entities. These major Air Force laboratories, or "super" laboratories as they were called, were: Armstrong Laboratory (San Antonio, TX), Phillips Laboratory (Albuquerque, NM), Rome Laboratory (Rome, NY), and Wright Laboratory (Dayton, OH). Organizationally, each laboratory was aligned with a given Product Center. For example, Wright Laboratory reported to the Aeronautical Systems Division, the predecessor to what is today known as the Aeronautical Systems Center (ASC). Like the other product centers, ASC in turn reported to the Air Force Material Command.

In response to several direct ed actions from Congress and the W hite House in the m id-1990s (see previous section on Di rected Actions and Im portant Studies for details), the Air Force initiated a plan to reconfigure and streamline its laboratory structure to produce a more integrated and cost-effective operation. This action ultimately led to the decision in 1996 to reorganize and consolidate res ources by establishing a single laboratory, AFRL. In addition to th e single laboratory concept, it was decided that AFRL woul d be commanded by a general officer and report directly to AFMC, just as the product centers, logistic centers and test centers did. These actions led to the creation of AFRL in 1997, which at that time consisted of the following technology directorates: Air Vehicles, Space Vehicles, Information, Munitions, Directed Energy, M aterials and Manufacturing, Sensors, Propulsion, and Hum an Effectiveness. The Air Force Office of Scientific Research, which supports research in academia, also reported to AFRL.

Biography
Russo's background includes an undergraduate degree in mechanical engineering and a PhD in m etallurgical engineering. His prof essional career was centered at W right-Patterson Air Force Base (W PAFB), home of Wright Laboratory, where he began his career as a m aterials scientist. In 1989, fo llowing several years in various m anagement and leadership positions at W PAFB, he became head of the Materials L aboratory. In the mid-90's, the leadersh ip within the Air For ce became very interes ted in the idea of reorganizing the infrastructure to incre ase integration and reduce the cost of it s laboratories. Improved efficiency was also an im portant goal, as a large potential manpower reduction was on the horizon. Russo's m anagerial skills attracted the attention of those responsible for this initiative, and he became the head of a transfor mation team responsible for the concept of converting the four super la boratories within the Air F orce (of which W right Laboratory was one) into a single laboratory (eventually known as AFRL). Following the successful form ation of AFRL, Russo moved to the Aeronautical Systems Center, where he became Executive Director. This center, one of four Air Force Product Centers, is responsible for the desi gn, development, and acquisition of aerospace weapon systems.

In total, Russo spent approxim ately 40 years within the Air Force technology and acquisition communities in one cap acity or another. Through out his long career with the

Air Force, Russo was very pro active in his managem ent style, looking for new w ays to improve his and other Air Forc e organizations. Russo felt strongly about leadership training, leading him to install an organizati onal development office within the Materials Laboratory during his tenure as its leader. When AFRL was established, a sim ilar office was established within the headquarters and at each of the technical directorates. At ASC, with a m uch broader set of responsibilities, Russo continued to inno vate in th is area, establishing an in-house cour se that em phasized leadership training. The course was designed and taught by senior AS C leaders. Russo's philosophy was one of "leaders teaching leadership." The course was very popular, and received high approval ratings. Russo expressed much pride in th is achievement. It should be noted that Russo retired from government service in 2004, but has conti nued his strong interest in the leadership area today from his position as the Presiden t and CEO o f Growing Splendid Leaders, LLC.

Interview Highlights
Regarding the AFRL technical program s, Russo's views reflected m any of the sam e thoughts that were expressed by Lyons and Coffey. To ac hieve world-class laboratory status, AFRL m ust strive to h ire world cla ss scientists and engineers, and at the same time provide these people with world class equipment and facilitie s. A strong cadre of world class scientists is essential to conducting quality in-house research, which in turn is central to building the required AFRL core competencies that support the Air Force mission. To m aintain the cadre of top resear ch talent in som e technical areas, Russo turned to regional universities and research firms to augm ent the in-house research staff. His approach was to contract with the un iversities and research f irms to pro vide individuals with the appropria te background that could comple ment and assist with S&T efforts at the in-house laboratories. This GOCA (government-owned, contractor assisted) approach worked well in all of the key in -house materials programs and the additional scientists working with the in -house scientific staff m ade significant contributions to the quality of the program.

Russo believed it was important to continually evaluate the AFRL technical program. To provide the AFRL leadership with a director ate-by-directorate assessment, Russo favored utilizing the Air Force Scientific Adviso ry Board (SAB). This group, m ade up of technical experts, addressed th e technical quality of each directorate's program, while participants from the Ai r Force Product Cent ers addressed the relevance aspects. T hese evaluations not only provided assessm ent of each directorate, but when com bined, assisted in obtain ing the AFRL m acro-picture. While directing the materials S&T program, Russo employed an ad hoc peer revi ew group to assess the technical quality of specific programs. He utilized na tionally recognized subject matter experts to review the program at the program elem ent level. This provided a more in-depth evaluation of technical quality than the di rectorate-by-directorate review of the SAB / Product Center. With the feedback from the peer review gr oup, Russo was able to fine tune the AFRL advanced materials program.

At ASC, Russo became m ore involved in th e transitioning of technolo gy, and realized that it was a m ore difficult challenge than he had originally thought, m ainly because the system program offices were not funded to integrate the products of the advanced development programs into their systems. To tackle the challenge, he helped institute an approach that coupled the technology needs of the system program office and the prime contractor with the laboratory funding proposals, mainly those in the advanced development areas. To aid in the process, the Air Force user commands were involved with evaluating the proposed projects. Upon evaluation, the projects were placed in one of three categories based on user needs, namely: user will provide transition funding, user will seek transition funding, and user not interested in seeking transition funding. To complete the cycle, approved projects were then assured of user-command support in the out-year budgetary process, an important requirement if the technology was intended for use by prime contractors. This process worked well and resulted in very favorable comments from industry, the user command, and the AFRL principal investigators.

Russo was also concerned about some perceptions and directions of the Air Force technology development program. From a quality viewpoint, the current and prior years' program have been very successful and have met Air Force needs. For the future, Russo hopes the trend continues. He emphasized, however, that care must be taken not to place too much emphasis on efficiency and expediency at the expense of technical quality and long term vision. If this change in emphasis comes to pass, Russo believes that the AFRL technology development program could be "headed in the wrong direction." While some in the private sector may not agree, it is the job of the in-house laboratory to act as an honest broker in the Air Force weapon system acquisition process. Without the underpinning of a quality S&T program, this role could be in jeopardy.

Russo also expressed concerns over some personnel related issues. Traditionally, AFRL has benefited from a strong cadre of technical managers who have had a long standing commitment to, and involvement with, AFRL. The trend toward hiring managers who lack years of experience at AFRL could negatively impact this cadre. Russo's concern was heightened by the Air Force leadership's interest in implementing a "geographical mobility model," whereby frequent reassignments of senior laboratory managers, such as those in the Senior Executive Service, would be mandatory. This approach mirrors the policy for active duty military officers, who are reassigned several times during the course of their careers.

Lastly, Russo expressed concern about the Project Reliance concept. The Services have worked hard to make this program a success. Russo believes that the most recent de-emphasis of the concept was not in the best interest of the overall DOD S&T program. The Project Reliance concept allowed the services to exchange ideas and share information regarding their technology development programs on a regular basis, and address issues of common concern. Russo believes that it is particularly important that Project Reliance succeed. He considers the idea of a single DOD corporate research laboratory or "purple laboratory," as it is sometimes called, to replace ARL, NRL, and AFRL, to be an unattractive alterative. Given the unique set of mission-related requirements for each service, Russo believes that it would be unrealistic to expect Air Force personnel, civilian or military, to have the same degree of confidence in a DOD-wide laboratory as they have today in AFRL.

Other Important Perspectives

Having heard from the former CRL executives, it is interesting to turn to those who have served in the Office of the Director, Defe nse Research and Engineering (DDR&E) and get their perspectives. The DDR&E is the office within DOD that oversees Service S&T activities in basic res earch, applied research, and advanced development. As part of this responsibility, it also addresses laboratory management issues.

Lance Davis was respo nsible for laborato ry management within DDR&E during the 1993–1998 timeframe, overlapping the tenures of Lyons, Coffey, and Ru sso as technical directors at their respective laboratories . Davis was interviewed in March 2008. This section discusses his thoughts, as well as those of Hans Mark, who was responsible for all of DDR&E in the time period (1998–2001) which followed Davis.

Lance Davis

Davis' background, including an undergraduate degree in metallurgical engineering and a PhD in engineer ing applied science, was ideally su ited for the DDR&E position. Following graduate school, Davi s served two years as a po stdoctoral student. He then joined Allied Che mical Company and continued his research in their corporate materials research laboratory. Following six years at the research bench, he became in succession group manager, materials research director and finally vice president for R&D. Davis ultimately spent about 25 years at Allied Chemical (now part of Honeywell Corporation). At Allied, Davis was part of a strong basic a nd applied research program. This research program led to products that provided Allied with a competitive advantage. One of the most important examples was m etallic glass research, which led to the development of low-loss magnetic m aterials that had dis tinctive advantages over more tr aditional materials. Things changed in 1992, when Allied's m anagement decided to enter o ther markets by aggressively acquiring other com panies. With these new acquisitions and the accompanying corporate culture change, there was less appreciation and support for the Allied research program, and the competitive advantages it provided.

In 1993, Davis m oved to the pub lic sector and took a p osition with DDR&E as the Director of Technology Transi tion. In a short period of ti me, his responsibilities were expanded to include laboratory m anagement. It was from this posit ion, as Director of Laboratory Management and Technology Transition, that Davis directed DOD's response to the congressionally m andated NDAA laboratory review of 1996. As noted previously, this response became known as Vision 21. Th e heart of the Vision 21 effort was an in-depth evaluation of all the Serv ice laboratories in s earch of ways to gain s ignificant improvements in operating efficiency.

The interview with Davis brought out severa l important thoughts rela ted to the in-house S&T laboratories. Davis knew the im portance of research from his tim e at Allied Chemical, and carried to DDR&E the belief that the in-h ouse laboratories play ed an important role within D OD. Davis saw this ro le as not only develo ping technology, but also serving other im portant Service needs, such as providing an h onest broker for technical opinions in the system acquisition cycle. Therefore, Davis believed that Vision 21 thrust should not be directed toward closing any of the laboratories, but rather towards developing an approach that would lead to reduced infrastructure and lower operational costs. The approach he took was based on a three-pronged effort based on reduction, restructure, and revitalization of the in-house laboratories. To accomplish the objectives, he had each Service form a Vision 21 team. The output of all three service teams was integrated under Davis' direction.

The Vision 21 report was completed and sent to Congress in April 1996.[11] It is important to note that the Vision 21 effort provided the impetus for the Air Force to move forward with a single laboratory concept (AFRL) as a means of achieving reduced operational costs. AFRL was activated in 1997. Highlights from the AFRL chronology, including the role of Vision 21, are shown in Appendix A.

Another effort that Davis strongly supported, and was the recipient of much his attention, was the Laboratory Quality Improvement Program. Under his leadership, the number of laboratories achieving "reinvention" status as provided by the 1995 NDAA increased significantly. Also, additional benefits were requested for laboratories covered under the LQIP demonstration project. These were:

- Authorization to exceed full-time equivalent authorization on a temporary basis
- Authorization to increase time to complete staffing actions
- Definition of high grades within the pay-banding system.

Davis retired from government service in 1998 and is currently Executive Officer of the National Academy of Engineering.

Hans Mark
Hans Mark has both an undergraduate degree and a PhD in physics. He has held research and faculty appointments at MIT and the University of California-Berkeley. Prior to heading DDR&E in 1998, Mark was Chancellor of the University Of Texas and served in previous administrations as Secretary of the Air Force and Deputy NASA Administrator.

Mark's views, published in the open literature and based on an examination of several government technology development laboratories (NRL, Oak Ridge National Laboratory, and NIST), [12] are pertinent to the subject of this paper. He highlights several important characteristics that were exhibited by these successful government technology development laboratories. Near the top of the list was basic research. He noted that performing basic research is essential to the work of the laboratory. It not only supports the laboratory's long-term mission, but it enables exploration into technical areas that are contiguous to the laboratory's mission, in case such areas may later become relevant.

---

[11] Department of Defense, *Vision 21*.
[12] Hans Mark and Arnold Levine, *The Management of Research Institutions* (Washington, DC: National Aeronautics and Space Administration, 1984).

Thus, a commitment to basic research enables the laboratory to move into new areas and to "diversify" as Mark put it, while sustaining the ability to conduct current programs.

In addition to conducting basic research and successfully diversifying its knowledge base, Mark notes the following important characteristics of successful government laboratories:

- Leaders of successful technology development laboratories must be responsible for the following priorities:
    o The primacy of diversification in decisions on the relevance of a technical area to current or future work pursued in the laboratory
    o Supporting basic research by: encouragement of scientists to publish their work in professional journals, collaboration with university researchers, sponsorship of interchanges of personnel between laboratory divisions as well as with the private sector, and sponsorship of improved instrumentation
    o Maintaining the proper balance between basic research and the more applied laboratory programs, including managing the internal perception that one group is gaining favor at the other's expense
    o Sponsoring discretionary research
    o Evaluating the technical quality of the work done by the laboratory and deciding how much of the laboratory's manpower can be devoted to customer funded programs
    o Maintaining the vitality of the laboratory by changing assignments, identifying candidates for management positions, instituting leave programs for professional development, and retraining staff when necessary
- The principal role of the laboratory is to strengthen the research and engineering base of new technologies, rather to serve as managers of large systems or to develop new products. That is not to say that occasionally a technology laboratory should not undertake a project if it has the technical capability to do so. Such an effort may provide the technology base for follow-on systems.
- The technology development laboratory is best administered as a number of loosely coupled units, rather than as a classical hierarchical system. Staff can then take responsibility for defining goals and communicating between the basic and applied sections of the organization.

Discussion and Concluding Remarks

While the thoughts of the laboratory execu   tives, Lyons, Coffey, and Russo, and the DDR&E officials, Davis, and Mark, were ex   pressed slightly diffe rently, threads of commonality are evident. These th reads were most apparent in areas related to b asic research, achieving world class status, peer re views, the role of gove rnment laboratories, and personnel m atters related to recruitment and retention of scientists and engineers. This section will discuss these areas and conclude with some recommendations.

There was general agreement among the executives that a basic research program was an essential underpinning to the technical success of their laboratories. In addition to putting a laboratory in a strong technical position to be an honest broker in the acquisition cycle, it was also essential for achieving world cla ss laboratory status, a goal sought by all three laboratory executives. Other fact ors contributing to world cl ass laboratory status were also mentioned. In addition to m odern equipment and facilities, it was im portant for scientists and engineers to:

- Have the opportunity to publish the results of their research in the open literature in refereed journals
- Be duly recognized by peers
- Earn a respectable salary
- Have the opportunity to transition findings to follow-on more applied programs
- Have the support of the acquisition and user elements of each Service. The latter requires a meaningful relationship with the customer, something that each executive worked hard to maintain.

To aid and guide the executives in m   aintaining a high quality technical program, a meaningful in-depth peer review process was necessary. There was agree ment among the executives that recognized external subject matter experts should be engaged to provide the review. It was noted that external subj ect matter experts could either be engaged by an external group such as the National Resear ch Council, or by the laboratory executives themselves. The selected subject m atter experts were to be not on ly leaders in their technical field, but independent thinkers as well.

Metrics such as the number of pat ents, publications, and number of PhD's were utilized, but in varying degree. ARL under John Lyons co llected a full range of metrics, but used only a selected num ber in managing the laboratory. Timothy Coffey found that a sm aller number would suffice for NRL.

Some concern was expressed by th e executives regarding the perception of the out side community toward government laboratories. The executives condemned the perception in some quarters that the in-house laboratories ex ecuted research of in sufficient quality, and that the Nation would be better served ha   ving academia or industry perform S&T for DOD. It was strongly believed by the executives that such a turn of events would clearly not be in the Nation's best interest.

In addition to the m any contributions that their laboratories have m ade to the Nation's defense, another important reason was given for the executives' strong beliefs that the in-house laboratories needed to be in the technology developm ent business. When it comes time to produce a given weapon s ystem, the pr ivate industry plays the dom inant role, providing most of the capabi lity to manufacture and field the weapon system. To win the production contract, the private sector competes to provide the government with technical proposals. The proposals in tu rn must be evalu ated. Acquisition officials who lack the special technical knowledge re quired to evaluate the perform ance claims of such proposals can turn to the subject m atter experts in the in-house labo ratories for unbiased reviews and recommendations. On m any occasions the technical experts reside in more than one Service laboratory, so that an increas ed measure of objectivity is possible. Thus, the in-house laboratories can be utilized as honest brokers to m ake the military smart buyers. It was strongly believ ed that hav ing this in -house capability significantly improved the likelihood that the DOD would avoid costly acquisition mistakes.

As to the reason behind the previously di scussed negative perception of the in-house laboratories, some interesting thoughts were provided by the executives. First, after the final production of weapons systems by industry, it is often very difficult to determine the origin of critical technologi es that end up in the fielde d weapon system . There is no marker on the weap on system that a given component was m ade possible by contributions from a given in-house laboratory. What is known is that the weapon system was produced by private industry. It is thus assumed by the public that industry was the source of all the critical technologies that made that weapo n system possible, when in fact it was, in m any cases, a team ing arrangement between the public and private sector that made the difference.

The executives also had som e important comments on scientist and engineer (S&E) personnel matters. Some discussion was centered on NSPS, the DOD-wide personnel system based on the principle of pay-for-per formance, which established four pay bands to replace the 15-step GS system. The executives would have preferred an S&E personnel system with the LQIP characteristics shown in Table 1. Under LQIP, laboratory management is empowered to a greater extent than with NSPS. The executives believed that while NSPS was a step back and much more bureaucratic, it was still workable. One executive stated that NSPS would not kill the laboratories, expr essing faith in their resiliency, and that the DOD laboratories would conduct successful S&T programs just as before.

Some discussion with the executives centered on related subjects. The first had to do with the idea of a "geog raphic mobility model" that appears to be under dis cussion in s ome quarters for very senior civi lian personnel. S uch systems could lead to the periodic movement of senior officials, such as Senior Executive S ervice members, from one location to another. Such mandatory moves would reflect the military model, and it was felt that such a system, if instituted, would hurt recru itment and retention, and in genera l would not be beneficial.

Concern was also expressed regarding the long-term outlook for DOD S&E workforce. This concern centered on outsourcing of th e in-house technology developm ent effort to the private sector. In add ition to the February 2008 interview, Coffe y has recen tly qualified his thoughts on the subject.[13] He has suggested that increased outsourcing could lead to an "accountability gap," and that th e DOD should not abdicate its role of m aking the key technical decisions involved in the acquisition of weapon system s. DOD i s the customer for these systems and is ultimately answerable to the public. However, as more and more of the DOD S&T is outsourced, the pr ivate sector workforce that is funded to do this wor k—as well as m uch of the adm inistration (planning, form ulation, review)— comes to play m ore of a dom inant role. This sets up a possible conf lict of interest, and raises the question of who is in charge of the DOD acquisition cycle, the governm ent or the private sector? Th e answer m ay have come from Congress,[14] which has m oved to limit the use of outsourcing, and with it the Lead System s Integrator approach that was utilized for the develop ment of such system s as the Arm y's Future Combat Syste m and the Navy's Submarine Combat System.

Coffey also notes that to meet these challenges, the DOD S&E workforce must keep pace with both the necessary talent and numbers, in comparison to the national S&E skill mix. Complicating matters is the fact that th e DOD S&E workforce continues to shrink, relative to the national S&E workforce. This s uggests that most of the future S&E talent may reside outside DOD. To o ffset these trends, Coffey proposes that the DOD S&E workforce, on average, center aroun d a fixed pe rcentage of the nation al workforce. Such an approach would go a long way toward addressing the above workforce issues.

Concluding Remarks and Recommendations
This paper has gathered the views of indi viduals who have collectively had m any years directing research laboratories within the Federal government. In particular, they directed S&T programs for the three m ilitary services at the Army Research La boratory, Naval Research Laboratory, and the Air Force Re search Laboratory. They have m ade the difficult executive decisions m entioned by Mark and de tailed in the previous section. These decisions played a significant role in shaping the in-house re search programs that formed the underpinnings of technology deve lopment for the Nation's weapons system s. As a result, Lyons, Coffey, and Russo get, in the author's opinion, high m arks for successfully meeting the challenges of their leadership positions.

With the above in m ind, it is felt that tapp ing the extensive wealth of knowledge and experience of these former laboratory executives in a single forum would pay significant dividends. The opportunity for them to inte ract with, and im part their advice and guidance to the current group of s enior S&T in-house laboratory m anagers, some of whom may be relatively new to their posi tion, would provide for a m ost beneficial exchange. Lastly, it m ay also be possible for these executives to participate in courses under consideration at the National Defense University.[15]

---

[13] Timothy Coffey, "Building the S&E Workforce for 2040," *Defense and Technology Paper 49* (Washington, DC: Center for Technology and National Security Policy, 2006).
[14] Elise Castelli, "US House Seeks Curbs on Contractors," *Defense News*, May 2008.
[15] Samuel Musa, private communication, December 2008.

Appendix A – Highlights from the AFRL Timeline[16]

1986	Packard Commission's blue ribbon study looks at ways to operate DOD in a more efficient and economical manner. David Packard, former Undersecretary of Defense, heads the study.
June 1986	Packard Commission issues its final report, *A Quest for Excellence*, proposing sweeping reforms to improve efficiency.
1986	President Reagan signs National Security Decision Directive 219, directing implementation of the major recommendations of the Packard Commission.
1986	President Reagan signs into law the Goldwater-Nichols Department of Defense Reorganization Act, considered the most significant defense-reform effort since 1947.
Late 1980s	With the fall of the Berlin Wall in 1989 and the dissolution of the Soviet Union in 1991, unpredictable regional conflicts replace the monolithic communist threat.
1989	President Bush hears complaints from Congressional representatives that the services are dragging their feet in supporting management reforms initiated by the Packard Commission and the Goldwater-Nichols Act.
February 1989	President Bush directs the Secretary of Defense to draft a plan to look at ways to improve management (with fewer employees) and organizational efficiency in DOD. The goal is to devise a strategy to implement sweeping reforms proposed in the Packard Commission's report.
June 1989	Secretary of Defense Cheney completes a major reorganization plan known as the Defense Management Review (DMR) that addresses ways to improve the defense procurement process. It urges military services to borrow/implement streamlined business practices used in the private sector.
July-September 1989	As part of the continual review process following the DMR report of June 1989, Secretary of Defense Cheney appoints special groups to investigate options for consolidating DOD functions, including laboratories.
October 1989	As a result of the special study groups, DMR Decision 922 strongly advises that the Pentagon consider merging all military laboratories directly under DOD.
December 1990	Thirteen Air Force laboratories become four.

---

[16] Excerpted from Robert Duffner, *Science and Technology—The Making of the Air Force Research Laboratory* (Maxwell Air Force Base, Alabama, November 2000).

October 1992	Consolidated Army Research Laboratory headquartered at Adelphi, Maryland, is formed.
November 1993	President Clinton announces a plan for an across-the-board review of all federal laboratories to streamline laboratory operations in view of projected decreases in federal R&D dollars.
May 1994	President Clinton issues a directive establishing the Interagency Federal Laboratory Review
May 1994-May 1995	At the direction of President Clinton, the National Science and Technology Council (NSTC) reviews the Nation's three largest laboratory systems operating within DOD, the Department of Energy, and the National Aeronautics and Space Administration.
Mid-1990's	The "planning studying, and assessing" phase of the laboratory structure ends. The Clinton administration, Congress, and DOD want the Air Force to take action and start reconfiguring the labs to produce an even leaner and more cost-effective R&D operation.
1995	DOD undertakes the Base Realignment and Closure (BRAC) process.
May 1995	Air Force leadership initiates a movement to produce a new Air Force vision for the 21st century. A study group that consists of senior military and civilian leaders is formed to come up with a more realistic vision that would be responsive to changing political conditions around the world.
May 1995	After a year of investigating how the DOD, DOE, and NASA laboratories operate, the NSTC submits its final report to President Clinton, indicating unanimous support for the legitimacy of laboratory functions but ample opportunity to improve management and cut redundancy.
February 1996	Congress passes the National Defense Authorization Act for fiscal year 1996. Section 277 of the legislation (Public Law 104–106) directs the Secretary of Defense to develop a 5-year plan to consolidate and restructure laboratories and test and evaluation centers assigned to DOD.
February 1996	President Clinton instructs the Secretary of Defense to submit a report to him, "detailing plans and schedules for downsizing the DOD laboratories."
April 1996	The Secretary of Defense delivers the *Vision 21* plan to Congress, which becomes the catalyst for the Air Force to move forward to completely revamp the laboratory system. The final *Vision 21* report combines outcomes from two studies—the DOD report compiled as a result of NSTC recommendations on laboratory restructuring, and the *Vision 21* DOD plan prepared in response to the

	Defense Authorization Act of 1996.
Late Spring 1996	The Air Force leadership proposes the formation of a single laboratory.
November 1996	The Secretary of the Air Force approves the single-laboratory concept.
January 1997	Dr. Russo is selected to serve as the single-laboratory transition director.
April 1997	By order of the Secretary of the Air Force, the Air Force Research Laboratory (AFRL) is activated.
July 1997	The AFRL organizational structure, including the consolidation of 22 technology directorates to 9, is announced.